Key Maths Skills Test Book

© C.R. Draper 2022

All rights reserved. No part of this book may be reproduced or transmitted in any form or by any means without written permission of the author.

Published: Warru Press, 2022

ISBN: 978-1-922819-01-7

Maths Test 1

1. In the number 43 875.6 what digit is in the thousands place? _____

2. 7 x 9 = _____

3. 534 + 376 + 92 = _____

4. 5963 – 785 = _____

5. A row of houses is numbered with the odd numbers between 31 and 59 inclusive. Which house is in the middle? _____

6. If 978 x 56 = 54 768, what is 54 768 ÷ 56? _____

7. Mrs Singh goes to the shops and buys 12 items at 50¢ each. She pays with a $10 note. How much change should she be given? _____

8. Mr Iaccuso takes his three children to the beach. He buys them each an ice cream for $1.20 and himself a drink for $2.50. How much does he spend? _____

9. Blake goes to the shops. Sticker sheets cost 30¢ each. How many sticker sheets can he buy with $2? _____

10. Each square, column and diagonal add up to the same number, in the number square below. What number does the question mark represent? _____

4		11
15		1
	?	12

Please use this page for working out.

Maths Test 2

1. In the number 87 645.3
   What digit is in the hundreds place?   _____

2. Deduct 7 thousandths from 23.45   _____

3. Of the numbers below, which is the biggest?   _____
   1.3   1.299   1.03   1.029

4. Round 5683 to the nearest hundred.   _____

5. Round 387.595 to the nearest hundredth.   _____

6. If Arosha spends $23.67 at the fruit and vegetable shop and $182.74
   at the supermarket, how much does she spend altogether? _____

7. Ruth goes into a shop and buys 9 sweets. She pays the shop assistant $1
   and gets 28¢ change. How much does each sweet cost?   _____

8. If pencils come in packs of 12, how many packs does Dominic need to buy
   if he needs 50 pencils?
                                                            _____

9. For a sports game, students are put into teams of 8. How many students
   do you need for seven teams?
                                                            _____

10. Ten DVDs fit into a rack and there are three racks in a box. How many DVDs
    can be stored in five boxes?   _____

Please use this page for working out.

Maths Test 3

1. Which number is the biggest?
   0.21,   0.201,   0.199,   0.2   _____

2. 3578 – 296 =   _____

3. 432.8 – 6.95 =   _____

4. If the temperature is -3°C overnight, but in the morning it has risen to
   5°C, how many degrees has it increased?   _____

5. Round 3.5 to the nearest whole number   _____

6. Kai goes to the shops. He has $2. Apples cost 60¢ and oranges cost 50¢.
   If he buys one apple, how many oranges can he buy?   _____

7. Abisha went to school on 181 days. There were 104 Saturdays and Sundays.
   There were 65 days on holiday. How many days was she away ill over the year?
   (The year had a total of 365 days).   _____

8. A supermarket shop came to $80.32. If they had $4.50 worth of savings,
   how much did they need to pay?   _____

9. If the average of four numbers is the total divided by 4, what is the
   average of 17, 26, 22, 17?   _____

10. Each row, column and diagonal add up to the same number, in the square
    below. Find A.   _____

13	6	11
A		
	14	7

Please use this page for working out.

Maths Test 4

1. 382 x 45   _____

2. 3654 ÷ 25   _____

3. What is the next number?   1,   3,   6,   10,   15   _____

4. If I think of a number, multiply by six and add five, the answer is 47.
   What was the number?   _____

5. One container holds 10 boxes. Each box holds 7 packets. How many
   packets in 9 containers?   _____

6. If 6 litres of juice cost $9.30, how much would 4 litres cost?   _____

7. If oranges are sold in bags of 8, how many oranges are there in 23 bags? _____

8. Priya's grandfather died in 2003 at the age of 76. In what year was he born? _____

9. When 16 is subtracted from half a certain number, the answer is 34.
   What is the original number?   _____

10. Deduct seven tenths from 1.45.   _____

Please use this page for working out.

Maths Test 5

1. Deduct 9 tenths from 3.66. _____

2. _____

    $15 \overline{)3547.5}$

3. 573.2 + 688.8 _____

4. $0.47 = \dfrac{?}{100}$ _____

5. Which is the largest number: 0.399, 0.401, 0.410, 0.409 _____

6. Round 3.895 to two decimal places _____

7. It reaches -7°C in the night but rises by 15°C by midday. What is the temperature at midday? _____

8. Calculate: 5 + 2 x (3 + 4) = _____

9. If Susan makes a batch of scones and gives half of them away to Arosha. She then gives four to Zachary and has eight left. How many did she make? _____

10. A postman delivers letters to the odd numbered houses on a street. If he delivers letters to the houses numbered 23 to 59 inclusive. How many houses does he deliver mail to on the street? _____

Please use this page for working out.

Maths Test 6

1. If the temperature on Wednesday was 7°C, and fell 12°C overnight, what temperature did it drop to? _____

2. If a straight line is 180°, what is the size of angle a?   145°/a   _____

3. 38.7 + 384 + 2.94 =   _____

4. If Michael's shopping comes to $137.64, but he then removes an item costing $12.76, what will his new total be? _____

5. What is the product of 153 and 46? _____

6. A business owner bought 1000 cups for $2630. How much would a single cup cost? _____

7. Work out as a decimal.   4)4398   _____

8. Chairs were stacked in stacks of 12. If there were 1152 chairs, how many stacks will there have been? _____

9. Denise bought 5 pens at 12p each, 8 pencils at 6p each, one pencil sharpener at 27p and 2 erasers at 15p each. How much did she spend altogether? _____

10. What is -8 – 4 ? _____

Please use this page for working out.

Maths Test 7

1. A carton of juice holds 6 cups of juice. How many cups in 234 cartons? _____

2. What is the value of the six in 4.762? _____

3. 3.9 – 0.04 _____

4. If Adhvika is four times her present age, she would be two years younger than her mother who is 30. How old is Adhvika? _____

5. $5 was shared among 4 girls and 6 boys. If each girl received 35¢ and the rest was shared equally between the boys, how much did each boy receive? _____

6. Uday bought 15 cartons of juice for $18.75. How much did each carton cost? _____

7. 12^2 _____

8. $\sqrt{64}$ _____

9. $32 + (2 \times 4) \times 2 =$ _____

10. Harshavi's watch is 6 minutes slow. At what time on her watch must she start her 9 minute walk to school if she has to be there by 8:50am? _____

Please use this page for working out.

Maths Test 8

1. Circle all the rectangular numbers below:
   21   23   25   27   29

2. Simplify $\frac{27}{63}$ _____

3. What is the highest common factor of 32 and 24? _____

4. The area of a triangle is half the base times height. What is the area of this triangle?
   Area (triangle) = $\frac{1}{2} b \times h$

   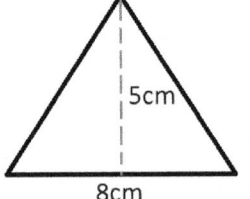

   the _____

5. What fraction of one hour is 45 minutes? _____

6. What is $\frac{4}{7}$ as a decimal to 2 decimal places? _____

7. What is the next number?   0.4   0.5   0.7   1.0   ? _____

8. Write $\frac{3}{4}$ and $\frac{5}{6}$ as fractions using their lowest common denominator. _____

9. Trains leave Reading station for Paddington every 10 minutes from 6:30a.m. What time does the fourth train leave? _____

10. A movie is 90 minutes long. They play it on TV but put 2 minutes of commercials after every 10 minutes of the programme. The movie is stopped in the middle for a 4 minute newsflash. If the movie starts at 7:00p.m., what time will it finish.
    (One hour is 60 minutes) _____

Please use this page for working out.

Maths Test 9

1. $3\frac{1}{2} + 1\frac{1}{4} =$ _____

2. $4\frac{1}{5} - \frac{4}{15} =$ _____

3. $\frac{7}{9} \times 12 =$ _____

4. What is the lowest common denominator of: $\frac{1}{3}, \frac{1}{4}$ and $\frac{1}{6}$? _____

5. Jeff ate $\frac{1}{3}$ of a pizza, Ciaran ate $\frac{1}{4}$ of the same pizza. How much pizza is left? _____

6. There are 1056 students and teachers in a school. If buses can carry 52 passengers, how many buses are needed to transport the whole school? _____

7. The volume of a cuboid is length x width x height. What is the volume of the cuboid shown? 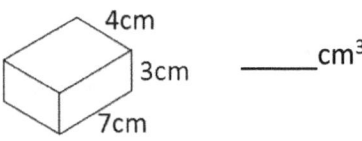 _____ cm³

8. Which number is prime? 25, 27, 57, 84, 97 _____

9. Calculate: $3^2 + (2+5) \times 3 =$ _____

10. On Wednesday the temperature dropped to -4°C overnight. At midday the temperature was 13°. How much had the temperature risen? _____

Please use this page for working out.

Maths Test 10

1. $\frac{3}{4} + \frac{1}{8} =$ _____

2. $\frac{5}{6} \div \frac{15}{36} =$ _____

3. $\frac{7}{11}$ of 55 = _____

4. If three identical apples were divided equally between four children, what fraction _____
   did each child receive?

5. Otylia drank $\frac{3}{4}$ of a can of lemonade and Shruthi drank $\frac{4}{5}$ of what remained. What _____
   fraction of the can of lemonade is left?

6. Antonio is $1\frac{3}{4}$ times younger than his brother Leo. If Antonio is 8, how old is Leo? _____

7. Alira is $1\frac{1}{2}$ times older than her younger brother Alys. If Alira is 12 how old is Alys? _____

8. The volume of a prism is: _____ cm^3
   the area of the base x height
   What is the volume of the hexagonal prism with a base of 7.3cm² and a
   height of 4cm?

9. When some oranges were shared, Priya received $\frac{7}{9}$ of them. She gave 5 to Magda and had 9 left.
   How many oranges were there before they were shared? _____

10. Races were run 10 minutes apart in a sports carnival. If the carnival started at 10:00a.m., what
    time did the fifth race start? _____

Please use this page for working out.

Maths Test 11

1. Convert 72% to a fraction. _____

2. Convert 57% to a decimal. _____

3. Find $\frac{5}{8}$ of 20 = _____

4. Calculate 9 + 2 x (5 – 4) = _____

5. A shop has a 25% off sale. How much would someone *save* if they buy jeans normally priced at $16 in the sale? _____

6. At a barn dance, apple pies were cut into eight. At the end of the evening there were 33 pieces left. Write how many apple pies were left as a mixed number. _____

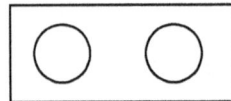

7. In the diagram above, two flower beds are surrounded by grass. If the area of each flower bed is $3\frac{3}{4}$m² and the whole are is $22\frac{1}{2}$ m², what is the area of the grass? _____

8. For a show, a theatre has sold $\frac{7}{8}$ of the available tickets. There are 90 tickets available at the door. What is the maximum capacity of the theatre? _____

9. Joel, Moses and Isaac share a small packet of sultanas. All 30 sultanas in the packet are eaten. Joel eats $\frac{2}{5}$ of them. Moses and Isaac eat the same number each. How many sultanas does Moses eat? _____

10. The area of a triangle is half the base x the height. What is the area of the triangle? _____

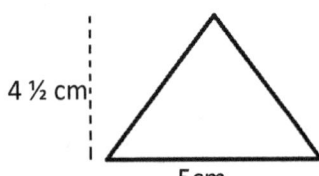

4 ½ cm

5cm

Please use this page for working out.

Maths Test 12

1. What is 24% of 150? _____

2. What is 35% as a fraction? _____

3. What is $1\frac{2}{5}$ as a percentage? _____

4. What is the highest common factor of 18, 54, 90? _____

5. Calculate $2^3 \times (3 + 2) + 5 =$ _____

6. Niamh had 24 counters. She gave $\frac{1}{3}$ to Jade. Jade then gave $\frac{1}{4}$ of what she had to Dominic. How many counters was Dominic given? _____

7. A furniture shop sells a desk at 40% profit for $84. How much did they purchase the desk for? _____

8. If there are 60 seconds in a minute and 60 minutes in an hour, how many seconds are there in one hour? _____

9. Ijeoma really likes a jacket that costs $160. She finds it in another shop for $120. What percentage does she save by buying it in the second shop? _____

10. Due to getting stuck in a traffic jam it takes Mr Smith 35 minutes to get to work instead of the normal 20 minutes. By what percentage was his journey increased? _____

Please use this page for working out.

Maths Test 13

1. 3.28 x 4.2   _____

2. 5.75 ÷ 0.25   _____

3. Write 0.625 as a fraction.   _____

4. Liam bought a computer game at a 20% off sale. If the price on the game is $12 how much did he pay for the game?   _____

5. Matthew is watching a video. He is $\frac{5}{7}$ of the way through. If the video is 35 minutes long, how much longer does the video have to go?   _____

6. In a class of 32 students $\frac{1}{8}$ of the students did not do their homework. How many students did their homework?   _____

7. Arosha bought a revision guide. The normal price is $15 but Arosha bought it for $10.50. What percentage did she receive off the cost of the book?   _____

8. On Tuesday the temperature was 11°C but fell by 16 degrees. What was the minimum temperature?   _____

9. If Serennita was 2 years older she would be a quarter of the age of her grandmother, who is 80. How old is Serennita?   _____

10. John takes his dog for a walk every morning. He increases the length of his walk from 4.8km to 5.4km. By what percentage has his walk increased?   _____

Please use this page for working out.

Maths Test 14

1. What is the mean of 18, 14, 25, 23? _____

2. What is the median of the numbers in question 1? _____

3. What is the range of the numbers in question 1? _____

4. Daniel was 16 in 2004. How old was he in 1990? _____

5. How many thousands are there in a $\frac{1}{4}$ of a million? _____

6. What is 75¢ as a % of $5? _____

7. What is the value of the 2 in 384.126? _____

8. A full rotation of a circle is 360°. What fraction of a full rotation is 90°? _____

9. Three out of every seven children in a class are girls. If there are 12 girls, how many children are there in the class? _____

10. Rena is at school from 9:00am to 3:00pm. What fraction of the day is she at school? _____

Please use this page for working out.

Maths Test 15

1. What is the next fraction? $1, \frac{1}{4}, \frac{1}{16}, \frac{1}{64},$ _____

2. 0.06 x 140 = _____

3. $12^2 - 11^2 =$ _____

4. 7.9km + 650m – 750m = _____ km

5. $1\frac{2}{5} \div 1\frac{13}{15} =$ _____

6. $30 \div (5^2 - 10) =$ _____

7. The middle of five consecutive odd numbers is 107. What is the first number? _____

8. 9 litres of olive oil cost $21.60. How much would 2 litres cost? _____

9. A shop sells a stereo for $169. If the profit was 30%, what was the cost price? _____

10. Reece goes on holiday to Ghana. He exchanges $500 for 2400 Ghanaian cedi. What is the cost in dollars of an item that cost 9.60 cedi? _____

Please use this page for working out.

Maths Test 16

1. 0.4 x 0.2 x 0.3 =   _____

2. $1\frac{2}{7} \div 1\frac{13}{14} =$ _____

3. 35% of 70 =   _____

4. How many hours in 3 days?   _____

5. $37^2 \div 37 =$ _____

6. The area of a parallelogram is the base times the perpendicular height. What is the area of the parallelogram?   _____

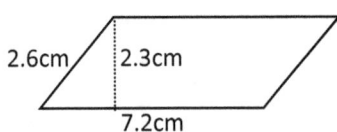

7. Karla bought a calculator for $8 in a 20% off sale. How much will the calculator cost after the sale has ended?   _____

8. 15kg of rice cost $18. How much would 25kg cost?   _____

9. Laura is swimming in a 1500m race. She is currently $\frac{2}{5}$ of the way through the race. How far has she yet to swim?   _____

10. Trains leave from Parmiter Hill train station every 7 minutes. If the first train leaves at 6:03am, what time does the sixth train leave?   _____

Please use this page for working out.

Maths Test 17

1. 45% of 180 =

2. What fraction of a week is 8 hours?

3. 3.2 ÷ 0.8 =

4. $(5 - 3^2)^2 - 2^3$ =

5. The area of a square is the length of a side squared. What is the **surface area** of a cube, with a length of 2cm?

6. A movie with a length of one hour 20 minutes was shown at 8:00pm Monday night. There were 25 minutes of ads and a 10 minute news break. At what time did the movie end?

7. Mirium is three times older than her brother. If Mirium is 16 years 3 months, how old is her brother?

8. Orla lives in England and goes on holidays to Europe. The exchange rate is £1 = €1.20. She buys a magazine for €3. What is the magazine worth in pounds?

9. The size of the family of a group of friends are: 3, 8. 4, 4, 3, 7, 6. What is the average family size?

10. What number is represented by the ? in the magic square below?

16	5	9	4
	11		14
3		6	15
13		?	1

Please use this page for working out.

Maths Test 18

1. 12% of 150 = _____

2. What fraction of an hour is 45 seconds? _____

3. $2\frac{5}{6} \div 1\frac{5}{12} =$ _____

4. What century was the year 1980 in? _____

5. $4^2 \div 2^3 =$ _____

6. 12.3kg + 250g + 3.8kg = _____

7. 4 pizzas cost $26. How much would 6 pizzas cost? _____
   (There is no special for buying multiple pizzas.)

8. Isaac has a 12 minute walk to the bus-stop. To get to school on time he needs to catch the 07:47 bus. What time does he need to leave home by his watch, if his watch is 4 minutes slow? _____

9. The formula for the area of a triangle is ½ base x height. What would be the area of the regular hexagon shown? _____

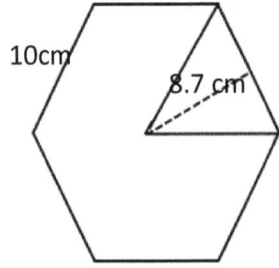

10. Wilette is a fast runner. She is keeping track of her times for the 100m. Over the last fortnight her recorded times are: 13.6s, 14.2s, 15.8s, 12.4s.
    What is her average time? _____

Please use this page for working out.

Maths Test 19

1. 300mg = _____ g

2. 4.2km = _____ m

3. What percentage of a day is 18 hours? _____

4. If it is 9am in London, what time is it in Tokyo? (Tokyo is GMT +9) _____

5. The temperature of the fridge section of a fridge-freezer is 4°C. The temperature in the freezer section is 25°C colder. What temperature is the freezer? _____

6. Salma catches the 09:02 train from Rockingham to Elizabeth Quay, Perth. The train takes 40 minutes. If her watch is 4 minutes slow, what time does she arrive at Elizabeth Quay station by her watch? _____

7. Ian is 3 times older than his brother. If Ian is 10 years, 3 months old, how old is his brother? _____

8. Mr Green buys four pairs of socks for an average of $1.50. He then buys a fifth pair for $4. What is the average cost for the five socks? _____

9. The piechart shows how Robert spent a day. How long did he spend doing homework? _____

   Sleep
   School
   Homework
   Eating
   Reading

10. A flapjack weights 40g. 15% of its weight is butter. How much butter is in the flapjack? _____

Please use this page for working out.

Maths Test 20

1. $4\frac{1}{6} \div \frac{5}{12} =$ _____

2. $0.2^2 =$ _____

3. Find the value of $7 \times 6 \div 14$ _____

4. Convert 0.04ℓ to mℓ _____

5. 3.2kg + 0.6kg + 50g = _____ kg

6. Saanya catches a bus to school. She catches the 07:43 bus and it takes her five minutes to walk to the bus-stop. If her watch is eight minutes slow, what time on her watch is the latest she can leave home? _____

7. Ric is making picture frames. A picture frame uses 70cm of wood. How many picture frames can he make from a 2.5m strip of wood? _____

8. The Seychelles are GMT+4. If it is 9:00pm in the Seychelles, what time is it in London? _____

9. A family go to a Dim Sum restaurant for lunch. The have 8 dishes with an average cost of $2.50. They order a ninth dish for $4.30. What is the average cost per dish now? _____

10. Kelsey is doing three tests. Each test is out of ten marks. She wants to achieve an average of $\frac{8}{10}$. If in the first two tests she gets $\frac{9}{10}$, what is the minimum score she can get in the third test to achieve her goal? _____

Please use this page for working out.

Maths Test 21

1. Convert 2.7m into mm. _____

2. Approximately how many centimetres are there in 5 inches? _____

3. $\sqrt{0.81}$ = _____

4. Write the factors of 40. _____

5. The temperature was -2°C and dropped by 5°C. What is the temperature now? _____

6. What is 6 out of thirty as a percentage? _____

7. Mrs Smith went to the supermarket. Her shop came to $83.27. She had $25 worth of vouchers which she used. How much did she need to pay? _____

8. Sasha's Mum is 3 times older that Sasha. Sophie is half the age of Sasha. If Sophie is 13 how old is Sasha's Mum? _____

9. The currency conversion between pounds and Australian dollars is £1 = $1.50. Carrie want to buy a pair of jeans on holidays on holiday which are $54.60. How much are they worth in pounds? _____

10. Jack had 90 beads. The beads are either green or yellow. Jack gave 2/3 of the beads to his friend. 2/3 of the beads Jack now has are green. How many yellow beads does Jack now have? _____

Please use this page for working out.

Maths Test 22

1. Convert 230g into kg. _____

2. $\dfrac{1.8}{0.9} =$ _____

3. $2^4 \times 5 =$ _____

4. What is the sum of the first five prime numbers?
   (Do not count 1 as a prime number). _____

5. 5 out of every 8 children in a class like maths. If 9 dislike maths, how many children are in the class? _____

6. What is the lowest common multiple of 6 and 8? _____

7. What is the shaded part of the circle as a fraction of the whole? _____

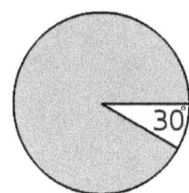

8. What is the size of angle a, in the quadrilateral below? _____

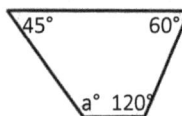

9. What is the name of the type of quadrilateral in question 8? _____

10. Joel buys a computer game for 12% off in a sale. If the computer game was $25 before the sale, how much does he save? _____

Please use this page for working out.

Maths Test 23

1. What is the next number?   7, 9, 12, 16,   _____

2. $\frac{3}{5} \div \frac{2}{5} =$ _____

3. What is the sum of 0.3, 0.05, 1.5 and 0.25?   _____

4. $4^3 - 2^5 =$ _____

5. 7 out of every 9 children in a class like reading. If 6 dislike reading, how many children are in the class?   _____

6. What is 6:30pm in 24 hour time?   _____

7. How many degrees does the minute hand of a clock move in five minutes?   _____

8. Three friends are playing a computer game for 4 hours 45 min. They take turns. What is the average time each friend spends playing the game?   _____

9. What is the area of the shape below?   _____

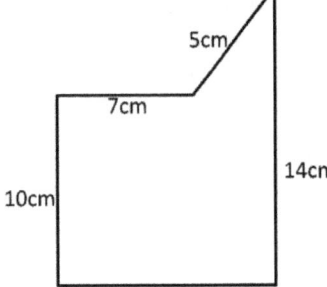

10. The rectangle below has an area of 4.5cm². What is the perimeter?   _____

Please use this page for working out.

Maths – Test 24

A shop has a 25% off sale.

1. A jacket was originally $84. How much would it cost in the sale? _____
2. Julie bought a skirt for $18 in the sale. How much would the skirt have cost before the sale? _____
3. The shop had this sign in the front.

   The shop is open for $9\frac{1}{2}$ hours a day. What time does it open? _____

4. If the day after tomorrow is Wednesday. What day was it the day before yesterday? _____

5. Toyin is having friends over, so is doubling her recipe. The recipe asks for $\frac{3}{4}$ cup of flour. How much should Toyin use? _____

6. A rectangle has a length of 3.2cm and a width of 3cm. What is its area? _____

7. A square has an area of 49cm². What is the length of a side? _____

8. Here is a diagram of a garden with grass around circular garden beds.

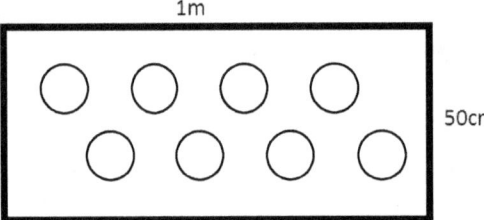

   Each circle has an area of 120cm².
   What is the area of grass around the flower beds (in cm²)? _____

9. Mrs Martin went to the shops and bought $96.20 worth of groceries. At the till she added a mobile phone top-up card costing $10. She had a $15 discount voucher for her shop. How much did she pay? _____

10. Two trains leave Paddington. One leaves every 6 minutes and one leaves every 9 minutes. If they both leave at 8:00a.m. When is the next time they leave at the same time? _____

Please use this page for working out.

Maths Test 25

1. $(0.5 \times 0.6)^2$ _____

2. $3.86 \times 7 \div 7 =$ _____

3. Put these in order of size, starting from the smallest. $37\%, \frac{1}{3}, \frac{2}{5}, 0.34$

   _____, _____, _____, _____

4. What is 35¢ as a percentage of $7? _____

5. If the perimeter of a rectangle is 20cm and the width is 4cm, what is the area of the rectangle? _____

6. What will 8 packs of muesli bars cost, if 20 cost $30? _____

7. Six more than a number is the same as the product of four and seven. What is the number? _____

8. A television programme lasted 1 hour and 40 minutes. It started at 6:15pm and there was a 15 minute news break. What time did the programme finish? _____

9. If it takes 8 people 12 days to build a shed, how long would it take 6 people? _____

10. 5 out of every 8 children in a class like creative writing. If 9 students dislike creative writing how many students are there in the class? _____

Please use this page for working out.

Maths Test 26

1. Subtract $\frac{3}{4}$ from 4.5 _____

2. What is 12 minutes as a percentage of an hour? _____

3. What is the name of the shape below? _____

4. What is the volume of the shape below? _____

5. If a square has an area of 36cm², what is the length of a side? _____

6. When $\frac{2}{5}$ of a certain number is reduced by 19, the result is 41. What is the original number? _____

7. What is a quarter past midnight on a 24 hour clock? _____

8. What number is half way between -7 and +9? _____

9. The average age of three siblings is 14. If the oldest is 22 and the youngest are twins, how old are the twins? _____

10. In assembly Kabir is in the middle of a row of students. He is 17th from the end. How many students are in the whole row? _____

Please use this page for working out.

Maths Test 27

1. $(4 + 5)^2 - (3^2 + 5^2) \times 2 =$ _____

2. $7^2 \div 14 =$ _____

3. What is $\frac{1}{3}$ of $\frac{7}{8}$? _____

4. What is the size of angle A? _____

   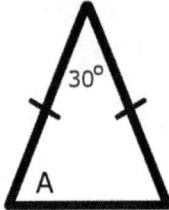

5. 186 children in key stage one at Achieve Primary School are right handed. 7% are left handed. How many children are left handed? _____

6. There are 60 stickers to a sheet. 48 are stars, the rest are dots. What is the ratio of stars to dots? _____

7. A picture frame has the ratio of 7:4. If it has a length of 21cm what is its width? _____

8. Rafael has 120 stickers. If he keeps 36 and then shares the rest equally among four friends, how many does each friend get? _____

9. There are a lot of posts lying on the ground and Amanda uses them to measure the length of the rectangle below. The rectangle has a length of 3 posts. What is the length of each post? _____

   Area = 90m²   5m

10. In a row of houses, the milkman delivers milk every third day and the baker delivers bread every fourth day. If they both meet on the third of May, what will be the date next time they deliver on the same day? _____

Please use this page for working out.

Maths Test 28

1. $\frac{1}{2}$ of $\frac{5}{6}$ = _____

2. 367 x 2.3 = _____

3. If A=2, B=5. C=7, D=9
   Then AB = 10
   What does CD= _____

On any given day in May, the probability that the pollen count will be high is 0.6.

4. What is the probability that on the 5th of May the pollen count is not high? _____

5. What is the probability that it will be high on the 5th, 6th and 7th of May? _____

6. A bag contains 5 blue balls, 3 red balls and 7 green balls. If Ian pulls out a ball from the bag. What is the probability that the ball will be blue? _____

7. Peter makes some flapjacks. The recipe asks for $1\frac{3}{4}$ cups of oats. If he is doubling the recipe, how many cups does he need? _____

8. Kumar buys 13 pairs of socks for $19.50. Pavel buys 7 pairs of socks. If the cost of each pair of socks is the same, work out how much Pavel paid for his seven pairs. _____

9. Ariana has 6 times more biscuits that Bertie but only three times more than Caitlyn. If there are 36 biscuits altogether, how many biscuits does Caitlyn have? _____

10. If Rikaz drives at 20mph for 90 minutes, how far will he travel? _____

Please use this page for working out.

Maths Test 29

1. Write these in order of size. $0.2, \frac{1}{3}, \frac{5}{6}, 0.4$ _____

2. $0.4^2 =$ _____

3. Calculate the size of angle A. _____

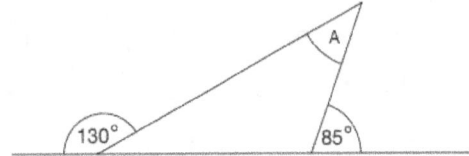

4. How many vertices does a triangular prism have? _____

5. 5.7km + 3.8km + 20m = _____ km

6. 5 + -9 = _____

7. Arosha has a bag of counters. There are 3 blue, 4 red, 5 green and 8 black. Which colour completes this sentence? The probability that it will be _____ is 0.2.

8. Samson won a million pounds in Lotto. He spent $10000 on a holiday. He spent half of the rest on a house. How much did his house cost? _____

9. The average height of a group of three friends is 1.5m. A fourth person joins the group. The average height of the group is now 1.55m. What is the height of the new group member? _____

10. Avery buys a jumbo pack of Apple and Raspberry flavoured sweets. $\frac{3}{8}$ of the sweets are apple flavoured and the rest are raspberry. If there are 15 apple sweets how many sweets are there altogether? _____

Please use this page for working out.

Maths Test 30

1. What is the mean of 54, 36, 53, 47? _____

2. What is the median of the numbers in question 1? _____

3. Aidan was 70 in 2013. In what year was he born? _____

4. What is 20 less than 2 million? _____

5. How many thousands are there in a $\frac{1}{4}$ of a million? _____

6. Xiou bought a jacket in a 40% off sale for $120. How much would the jacket of cost if he had bought it the day before the sale _____

7. How many degrees does the minute hand of a clock turn from 9:40am to 10:05am? _____

8. Ilya spends 8 hours at an activity centre. She spends 45 minutes eating lunch and the rest of the time doing three activities. How long does she spend on each activity if she spends the same amount of time on each activity? _____

9. Jacob thinks of a number, multiplies by eight and adds 12. Half of his final number is 50. What number did he start with? _____

10. If Sonya faces East and turns 135° anticlockwise, in what direction will she face? _____

Please use this page for working out.

Maths Test 31

1. $(3^2 + 9^2) \times (3+9)^2 =$ _____

2. $12x + 13 = 253$ Solve for x. _____

3. Ursula bought a box of books for $175. She sold each of the 25 books for $10.50. What is her percentage profit? _____

4. Using the graph below. How many pounds are there in 45 kilograms?? _____

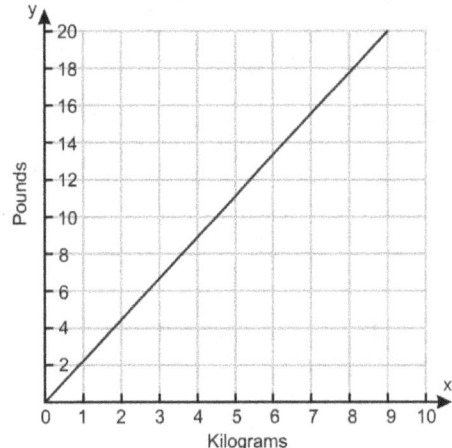

5. Stefan had a stamp collection, containing 180 stamps from around the world. If he kept a quarter of the stamps for himself and then shared the rest out equally between 5 friends. How many stamps does each friend receive? _____

6. The coins below are in a line. What is the diameter of the small coins? _____

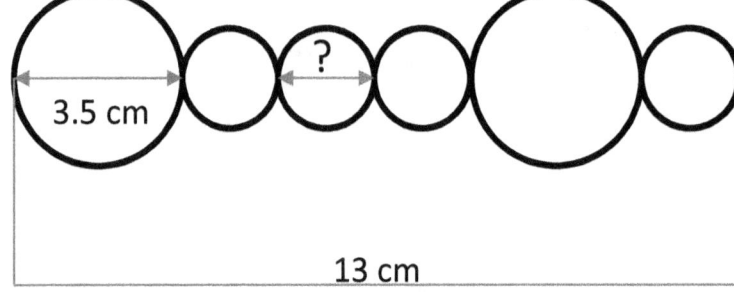

7. Ryan puts some water in a measuring cylinder, then drops in a rock.
   What is the volume of the rock? _____ ml

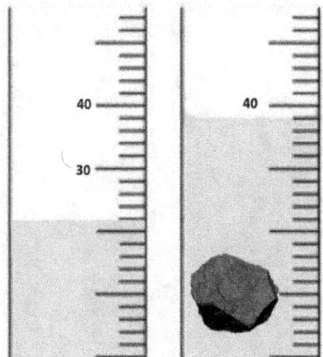

8. Flynn has a large pencil case. He has three pencils for every two felt-tips. If he has a 12 pencils, how many writing implements does he have in his pencil case? _____
   (assume all writing implements are either pencils or felt-tips).

9. A rectangle has corners at A, B, C and D. Point E is in the middle of the rectangle? What are the coordinates of point E? _____

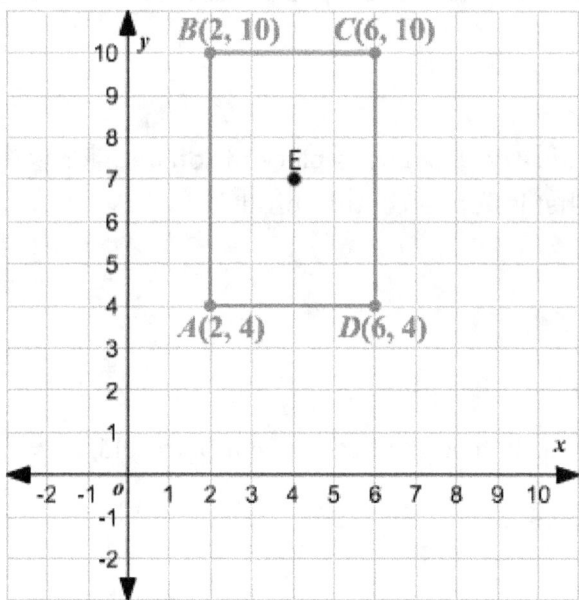

10. If the rectangle is moved so that the coordinates of A are (-1, -3) what are the new coordinates for point C? _____

Maths Test 32

1. 3480 ÷ 15 = _____

2. A = 42, B=6, C=27, D=23, E=15 _____
   Calculate and give you answer as a letter: 2B + E =

3. Solve for a. $\frac{12+a}{3} = 5$ _____

4. Jagmar caught the 7:55am train to work. The train takes half an hour. It takes her 15 minutes to walk from the station to work. However, she stopped to help someone who needed first aid for 23 minutes. If she is meant to be at work at 9:00am, how late was she? _____

5. Doris was catering for a large function of 1000 people. She wants to buy apple pies and custard tarts in the ratio 3:2. Each pie and tart is to be cut into eight pieces, and she needs to buy enough for each person to have one piece of desert. How many apple pies does she need to buy? _____

6. Hannah works in the computer industry. She buys a piece of software for $45 and sells it to a client for $81. What is her percentage profit? _____

7. Will the year 2020 be a leap year? _____

8. Devin is seven times his sister's age. If he is 10 years and 6 months old, how old is his sister? _____

9. What is the area of a rectangle with a length of 3.2 cm and a width of 1.5 cm? _____

10. Paul has 6 games, 15 puzzles and 3 teddy bears. If he can choose one game, one puzzle and one bear to take away on holiday, how many possible combinations does he have to choose from? _____

Please use this page for working out.

Maths Test 33

1. How many thousands are there in a quarter of a million?   _____

2. $\dfrac{5}{9} \div 8\dfrac{1}{3} =$ _____

3. Solve for m.   6m + 11 = 14   _____

4. Jessica has an average of 76% for 4 exams. If she has two more exams what is the largest average she can get for all of her exams?   _____

5. Jacqui and Anne grow tomato plants. The ratio of tomato plants that Jacqui has to Anne is 5:7. If Jacqui has 20 tomato plants, how many more tomato plants does Anne have?   _____

6. Mrs Browne goes to the shop. She spends $57.80. However, she has a 15% off voucher. How much does she save?   _____

7. The perimeter of a rectangle is 12cm. One side is 3.5cm. What is the area of the rectangle?   _____

8. If a TV station starts a 90 minute movie at 7:55pm and it finishes at 9:45pm, how many minutes of advertisements are there?   _____

9. What is the name of the shape below?   _____

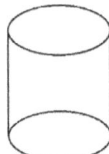

10. Grainne has a part time job. On Saturday, she started work at 10:30am and finished at 4:30pm. If she is paid at a rate of $5.20/hour, how much did she earn?_____

Please use this page for working out.

Maths Test 34

1. If 37 x 63 = 2331, what is 3.7 x 0.63? _____

2. Solve for a. $3\frac{a}{7} + 5 = 14$ _____

3. For the following number sequence, what is the 40th term? 45, 48, 51, 54 _____

4. If 4/5 of a class like reading and 6 students dislike reading. How many students in the class? _____

5. Serena thinks of a number, doubles it, adds 6 and divides by 3. If the answer is 4, what was her original number? _____

6. The average amount of money spent by five girls on a shopping trip is $25. The average spent by four of the girls is $24. How much did the fifth girl spend? _____

7. Find the size of angle a.

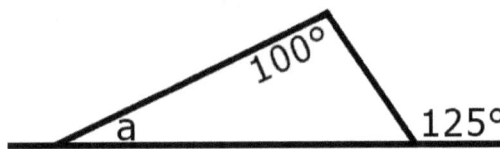

_____

8. What is the sum of the prime numbers between 50 and 70? _____

9. 150 children in a large primary school sat the year six SATS exam. 70% achieved level five of higher. How many children did not achieve level 5? _____

10. A group of friends record how many vehicles pass a road in an hour. 20% of the vehicles were buses. There were twice as many buses as vans. How many buses did they record? _____

Vehicle	Number
Car	120
Bus	
Van	
Motorcycle	14
Bicycle	6

Please use this page for working out.

Maths Test 35

1. Eight friends win $70 in a competition. If they share the prize money equally how much does each friend get? _____

2. Terrence buys 12 muffins for $5, then sells them for $1 each. What percentage profit does he make? _____

3. Nikhita drops a ball from a window 42m above the ground and it bounced back two-thirds of the distance that it had fallen. What height did the ball bounce to? _____

4. Maya read one quarter of her book on Friday and half of the remaining pages on Saturday. If she has 90 pages left to read, how many pages are there in the book altogether? _____

5. The school coach normally arrives at the school at 3:40pm but was stuck in traffic for 17 minutes. If Joseph has a 25 minute coach trip from school and a 12 minute walk from the bus stop, what time did he get home that day? _____

The angles of the triangle below are x, x+20 and 2x.

Not drawn to scale

6. What type of triangle is it? _____

7. What is the size of the largest angle? _____

8. Diana is cooking dinner. If the recipe uses 360g of flour for four people, how much flour does Diana need to use for 7 people? _____

9. Six teams play in a backgammon competition. If each team plays each of the other teams exactly once, how many games are there? _____

10. Desmond counts in eights. If he started from a number less than ten and reaches 524. What number did he start counting from? _____

Please use this page for working out.

Answers

Test 1

1. 3
2. 63
3. 1002
4. 5178
5. 45
6. 978
7. $4
8. $6.10
9. 6
10. 7

Test 2

1. 6
2. 23.443
3. 1.3
4. 5700
5. 387.60 (accept 387.6)
6. 206.41
7. 8¢
8. 5
9. 56
10. 150

Test 3

1. 0.21
2. 3282
3. 425.85
4. 8°
5. 4
6. 2
7. 15
8. 75.82
9. 20.5
10. 8

Test 4

1. 17 190
2. 146.16
3. 21
4. 7
5. 630
6. $6.20
7. 184
8. 1927
9. 100
10. 0.75

Test 5

1. 2.76
2. 236.5
3. 1262
4. 47
5. 0.410
6. 3.90
7. 8°C
8. 19
9. 24
10. 19

Test 6

1. -5°C
2. 35°
3. 425.64
4. $124.88
5. 7038
6. $2.63
7. 1099.5
8. 96
9. $1.65
10. -12

Test 7

1. 1404
2. 6 hundredths
3. 3.86
4. 7
5. 60p
6. $1.25
7. 144
8. 8
9. 48
10. 8:35am

Test 8

1. 21, 25, 27
2. $\frac{3}{7}$
3. 8
4. 20cm³
5. ¾
6. 0.57
7. 1.4
8. $\frac{9}{12}, \frac{10}{12}$
9. 7:00a.m.
10. 8:50p.m.

Test 9

1. $4\frac{3}{4}$
2. $3\frac{14}{15}$
3. $9\frac{1}{3}$
4. 12
5. $\frac{5}{12}$
6. 21
7. 84cm³
8. 97
9. 30
10. 17°

Test 10

1. $\frac{7}{8}$
2. 2
3. 35
4. $\frac{3}{4}$
5. $\frac{1}{20}$
6. 14
7. 8
8. 29.2
9. 18
10. 10:40 a.m.

Test 11

1. $\frac{18}{25}$
2. 0.57
3. 12.5 or $12\frac{1}{2}$
4. 11
5. $4
6. $4\frac{1}{8}$
7. 15 m²
8. 720
9. 9
10. $11\frac{1}{4}$cm²

Test 12

1. 36
2. $\frac{7}{20}$
3. 140%
4. 18
5. 45
6. 2
7. $60
8. 3600
9. 25%
10. 75%

Test 13

1. 13.776
2. 23
3. $\frac{5}{8}$
4. $9.60
5. 10 minutes
6. 28
7. 30%
8. -5°C
9. 18
10. 12.5%

Test 14

1. 20
2. 20.5 or $20\frac{1}{2}$
3. 11
4. 2
5. 250
6. 15%
7. 2 hundredths
8. $\frac{1}{4}$
9. 28
10. $\frac{1}{4}$

Test 15

1. $\frac{1}{256}$
2. 8.4
3. 23
4. 7.8 km
5. $\frac{3}{4}$
6. 2
7. 103
8. $4.80
9. $130
10. $2 or $2.00

Test 16

1. 0.024
2. $\frac{2}{3}$
3. 24.5
4. 72 hours
5. 37
6. 16.56 cm²
7. $10
8. $30
9. 900 m
10. 6:38 am

Test 17

1. 81
2. $\frac{1}{21}$
3. 4
4. 8
5. 24cm²
6. 9:55pm
7. 5 years 5 months
8. £2.50
9. 5
10. 12

Test 18

1. 18
2. $\frac{1}{80}$
3. 2
4. 20th (century)
5. 2
6. 16.35 kg
7. $39
8. 07:31
9. 261 cm²
10. 14s

Test 19
1. 0.3g
2. 4200m
3. 75%
4. 6pm
5. -21°C
6. 09:38
7. 3 years 5 months
8. $2
9. 6 hours
10. 6g

Test 20
1. 10
2. 0.04
3. 3
4. 40ml
5. 3.85
6. 07:30
7. 3
8. 5:00pm
9. $2.70
10. $\frac{6}{10}$

Test 21
1. 2700mm
2. 12.5cm
3. 0.9
4. 1, 2, 4, 5, 8, 10, 20, 40
5. -7°C
6. 20%
7. $58.27
8. 78
9. £36.40
10. 10

Test 22
1. 0.23kg
2. 2
3. 80
4. 28
5. 24
6. 24
7. $\frac{11}{12}$
8. 135°
9. (irregular) trapezium
10. $3

Test 23
1. 21
2. $1\frac{1}{2}$ or 1.5
3. 2.1
4. 32
5. 27
6. 18:30
7. 30
8. 1h 35min
9. 106cm²
10. 9cm

Test 24
1. $63
2. $24
3. 8:00a.m.
4. Saturday
5. $1\frac{1}{2}$ cups
6. 9.6cm²
7. 7cm
8. 4040cm²
9. $91.20
10. 8:18 a.m.

Test 25
1. 0.09
2. 3.86
3. $\frac{1}{3}$, 0.34, 37% $\frac{2}{5}$
4. 5%
5. 24cm²
6. $12
7. 22
8. 8:10pm
9. 16 days
10. 24

Test 26
1. 3.75 or $3\frac{3}{4}$
2. 20%
3. Sphere
4. 1100 cm³
5. 6 cm
6. 150
7. 00:15
8. 1
9. 10
10. 33

Test 27
1. 13
2. 3.5
3. $\frac{7}{24}$
4. 75°
5. 14
6. 4:1
7. 12cm
8. 21
9. 6m
10. 15th May

Test 28

1. $\frac{5}{12}$
2. 844.1
3. 63
4. 0.4
5. 0.216
6. $\frac{1}{3}$
7. $3\frac{1}{2}$
8. $10.50
9. 8
10. 30 miles

Test 29

1. 0.2, $\frac{1}{3}$, 0.4, $\frac{5}{6}$
2. 0.16
3. 35°
4. 6
5. 9.52
6. -4
7. Red
8. $495 000
9. 1.7m
10. 40

Test 30

1. 47.5
2. 50
3. 1943
4. 1 999 980
5. 250
6. $200
7. 150°
8. 2 hours 25 min
9. 11
10. NW (north west)

Test 31

1. 12 960
2. 20
3. 50%
4. 100 lb
5. 27
6. 1.5cm
7. 16
8. 20
9. (4,7)
10. (3,3)

Test 32

1. 232
2. C
3. 3
4. 3 min
5. 75
6. 80%
7. yes
8. 1 year 6 months
9. 4.8 cm²
10. 270

Test 33

1. 250
2. $\frac{1}{15}$
3. 0.5 or $\frac{1}{2}$
4. 84%
5. 8
6. $8.67
7. 8.75cm²
8. 20 min
9. cylinder
10. $31.20

Test 34

1. 2.331
2. 42
3. 162
4. 30
5. 3
6. $29
7. 25°
8. 240
9. 45
10. 40

Test 35

1. $8.75
2. 140%
3. 28m
4. 240
5. 4:34pm
6. scalene
7. 80°
8. 630g
9. 15
10. 4